INTERNET DOPE IS MY LIFE

Will Smartech Kill a Future Generation?

G. Taboris Taylor

INTERNET DOPE IS MY LIFE

ALL RIGHTS RESERVED. No part of this book may be reproduced or transmitted for resale or use by any party without the full consent of Gregory Taboris Taylor, owner of Taboris Intelligence Asset Group, LLC. Any reproduction or transmission of this book, in any form or by any means, electronic or mechanical, including photocopying, recording, or by any information storage or retrieval system, is prohibited without express written permission from Mr. Gregory Taboris Taylor.

Photo cover Courtesy of Graphic Designer Donahue Johnson; Public Domain pictures, p. 2, 3, 12, 14, 41-44, 47, 49, 51, 52; Luke Benoit p.45

Copyright © 2019 by G. Taboris Taylor

Table of Contents

Introduction, 1

Survey: Electronic Dependency and Smartphone Addiction, p.12-13

The Internet Plague, Certified Smart People Universal Questions, p.1-10, 15, 26

Chatbot ai Revolution, Side Effects of Digital Drugs, p.11, 14, 15-18

American Dances, p.22; Change Creation & Innovation, p.40-42

Mindset of our Youth, p. 24-28, 34; Digital Booty p. 58

Gangstacrats n Rethuglicans Political Duopoly, p.21, 31, 41

Old School Bully true story rap, p.23

Social Media & Neighborhood Olympic Challenges, p.27-28

50 years of American Slang, p.29-31

Things we're never going to Agree on, p.31, 40-41

Kewl parenting 101, Maybe Junkie, World Wide Web, p.34-39, 60

Self Esteem, Emotional Intelligence, Millennial ThinQ4self Matrix, p.25, 32-40

Code name: Operation Self Hate Terminate & Eliminate, p. 57

Real Talk, p. 37-49, 58, 71

Fake Life, p.42-43

Quick Rhyme p.44

Internet Plaque Antidote, p.59-62

Glossary, p. 63-70

Introduction

Unbanked! Unbought! Unbossed! Caution! This information may potentially stimulate brain triggers for those worth barely 5 figures whose real life does not mirror what they project themselves to be in their fake world, trapped in the virtual fortress and is afraid to take that Mask off and **Keep it 100** in the **Real World**—Reader discretion is advised; so don't call the prayer police, the thought police and the Fun Police on me because I'm not like 99% of the people in the all of a sudden PC America walking around on eggshells censoring themselves and are afraid to speak their mind. In this ever-changing world and increasing automation, whether it is scan n go at Wally World, self-driving cars, flying motorcycles, a factory robot spilling a chemical in a warehouse, everyone looking down never up, disrespectful and spoiled ungrateful kids, those online chatbots answering your questions and responding to tweets and retweets or having armed guards at your local super market to keep an eye on those delicious tide pods and bath salts. How do we turn GangstaCrats and Rethuglicans back into Democrats and Republicans then back into Americans? How did people order food b4 value meal combos in 1992? Who's the gate keeper of your Pii? If the doctor diagnosed you with a stomach virus on Tuesday and asked you, what did you eat over the last 3 days and you say: well on yesterday I ate McDonald's for breakfast, Chik-fila for lunch, Roast Duck for dinner. On Sunday I had Waffle House for Breakfast, Chipotle for lunch and ate at Chili's for dinner. On Saturday I had **iHop** for Breakfast, some Gumbo and hot wings for lunch and a fried seafood platter for dinner at Red Lobster. Then it dawns on you that after all of that food back tracking and Starbucks every day; you don't have a clue what contributed to unintended consequences making you sick and furthermore; you used your debit card to make all of those purchases compromising your personal financial information—duh! Don't get lost in the sauce because what is discerning to me is that Bill Gates made 15 predictions in 1999 in his book, **Business @ the Speed of thought**; Outside of Washing and Drying your Clothes, people will carry around small devices that will allow them to Communicate worldwide digitally with literally all of the worlds knowledge at their fingertips; but now I got my phone in my hand and can't put it down, I know I need to stop but don't know how—thanks Steve Jobs! Why are several high-profile Silicon Valley renegades sounding alarms in increasingly ominous terms about what smartphones and social media do to the human brain? By 2025 will devices be operated by what we say and not by what we press. And parents thought kids playing post Office when no adults were around back in the day, being self-reliant or succumbing to someone with a big wad of cash was the biggest threat to kicking in that door of full potential stunting your personal development. Well I got news for ya! It isn't sir, podner or lil homie depending on Yo Neck of the woods: United Nations, the Vatican, Congress on the Hill, Park Avenue, Beverly Hills, Suburbs, the Neighborhood, back there in the cut, new and old projects, extended stay motels, back woods or up there in the Mountains. There are over 4,000 satellites at this very moment circling the earth watching us; over 700k drones in America and over 300 million Ghosts in the Sky aka surveillance cameras worldwide watching and listening to us everywhere: on our streets, at work, in restaurants, in schools, in our homes and hands. **The World Wide Web and The Internet** has changed and revolutionized the way we live and work for many of us both young and old, but like numerous modern inventions throughout time, its advantages haven't come without consequences. We all know and see these people or dare I say Cyborg like species looking down never up, who appear to have a severe Dependency or Electronic imbalance with the Internet, Social Media and Smartphones just about everywhere—let's all say it out Loud together: Baby, I luv ya but you have an Internet Addiction Disorder (**IAD**) called Smartphone Addiction. It's the pink elephant in the room right there alongside Race, Religion, LGBT and Politics and no one wants to have a difficult conversation and talk about a comprehensive solution to deal with what's going on with the mindset of 9-50-year-olds in America. Are these smart, intelligent, educated individuals who check their phone 100 times a day incapable of realizing that they are living in a Virtual World that is an alternate or parallel universe to the Real World and Social Media is not reality—their life now consists of 140 characters, all about the swag and drip drip, viral videos, streaming live, selfies, group texts, posts, finding humorous images on unsplash to post on Instagram and Facebook, tweets, retweets, emojis by their statements, likes and floating hearts to describe their mood? What do you cling to when Reality has left you nothing else? Why did Bill Gates, Mark Zuckerberg and the late Steve Jobs admit that they put restrictions on their own kid's usage of smartphones? Why has there been a rise in Mental Health issues and Depression since the proliferation of social media and powerful devices like smartphones? How much exposure to smartphones and social media is OK for the brain? Why are our young males afraid to make eye contact when introducing themselves or engaging in a conversation and females are not? Are Angels the Warriors for God? Why are young men making consent videos on college campuses with women prior to going out on a date? Can the over usage of smartphones and technology lead to an Addiction or a Digital and Electronic Dependency? How do we teach the next generation to Learn to Look up Again? Is Smart Tech making the next Generation

Dumber? Does all knowledge of Reality begin with an Experience and Ends in it? When Tweety Bird said, "I thought I saw a Putty Cat", is that a Tweet? Has social media ruined your relationship? Is your phone ruining your Mood? Is America still Tone Death? Are technologists in Silicon Valley becoming more mindful of the dangerous effects significant screen time can have and aren't allowing their own children to partake in it? Will Technology Kill a future Generation? Why are parents in the tech industry banning their children and nannies from using smartphones and giving them dumb phones? How much longer will Screens be seen as a learning tool for children since the risks for addiction and stunting personal development seem high? Will we need voice fraud protection or smart lock integration in 2020 Siri, Alexa, Bixby and Cortana? Are we living in an age of Apocalypse? Are we living in a Time where the Economy and Political System can't be fixed by any human? What did Mr. Theodore Parker and Dr. Martin Luther King Jr. mean when they said the "arc of the moral universe is long, but it bends toward justice?" What do you do when you start believing things that aren't true? What's worse, a Bold Face lie or a Half-Truth? What happened to ladies getting summertime Fine in America? What happened to national letter writing day? Is it offensive when a millennial says U straight? Are artists and entertainers engineers of the soul? How do you get notified each time someone searches for you on Google? Does taking a 10 min walk without your phone and earbuds on really stimulate Brain Activity? Is exercise and laughter the best Medicine? Would senseless murders of unarmed Blacks end if every black male under the age of 45 was insured for $5 million? Does Siri say I'll holler when minorities get pulled over by the police? Can a lone voice raise the voices of many? Why do millennials think getting a good deal on electronics on Black Friday is an Investment? Are we Using Technology or Is it Using Us? Is there a Science to succeeding with people? Are you living an Insta-Lie? When the winds of change are upon us would U rather be a soldier, warrior or leader? When will your family eat its final meal together? Would u sell your only child for a billion dollars? What do you do when the weakest sibling becomes the strongest and is now the bully? Why do winners worry about winning and Losers worry about Winners? Are you going to believe me or your lying ass Eye's? Will Americans riot in the streets when gas is $8 a gallon? Will the Universe unfriend us and make us extinct like the dinosaur? Where in the ham sandwich does the Internet, Google, Siri and Alexa get all that information from anyway? What do you call a duck named nuck or a buck named chuck in November? What is the Secret to Life? Is there a stigma attached to IAD or smartphone addiction? Is criticism a friend or foe? When someone criticizes you do you say to yourself you just wait I'm gonna prove you wrong or do you get mad and become vengeful? For instance, were you ever told you're just like your no-good father or mother or I wish you could be more like your sister or brother because their someone to be like? Is this Generations' mindset hard wired for Apocalyptic Thinking? Outside of the Library, Encyclopedias' Britannica and World Book, the Internet, Google, Siri and Alexa of my day when I graduated from high school in 1985 were Watson and people who actually cared about you: Grandparents, Parents, Uncles, Aunts, Big Brothers and Sisters, Cousins, Neighbors, Teachers and Leaders in the Community were the sources of information and wisdom. Don't pay attention to the man or woman behind the curtain. Why are the only people that will tell you the Truth about anything are the Angry, Kids and Drunks? But I'm neither! I'm a disrupter of modern-day television like Netflix—that prophet of walk around sense and spark in a room full of Gasoline and Super Thermite. I'm an equal opportunity offender of Democrats and Republicans (left-right-liberal-conservative-GOP-progressive-log cabin-jackasses-elephants); professional politicians, bitter men and women, lazy men and women, jealous men and women, broke before payday men and women, dish it out but can't take it men and women, dog bigots, food bigots, fake bullies, internet trolls, need a 1977 switch off a tree time out for ungrateful disrespectful children, video game dope fiends, internet dope fiends, lazy grown up kids, poverty pimps, twitter thugs, digital pimps, vanity and self-esteem pimps and people who spend their day taking selfies and hating instead of focusing on checking and savings. What does it mean when people say to be perfectly Honest with you? Does it mean just because you're a Christian and profess you Love God you want lie to or on me? Skin color prejudice is so pathetic it's almost quaint! If you're a fool and you go to college then you're an educated fool; if you're an uneducated fool who drinks red bull, monster and love them Juul vape pins, you're a New Damn Fool! Is riding around on a moped/motorbike with earbuds on just Dumb? Are wearing headphones or earbuds prohibited while driving a vehicle? This generation for some reason can't get enough of Stow Bought Sense; Every time they Learn Something they Pay 4 it literally! lol. Our youth are screaming out for help and nobody's listening so, I think I better call the General Data Protection Regulation in **Europe** since no one in America gives a Dam about the **Youth's physical and emotional wellness**—the motto in America is just give that kid a smartphone, cookie and earbuds—that'll keep'em quiet so Cool Mom and Cool Dad can turn up on Social Media or play candy krush—I remember years ago these same people were playing angry birds. Better yet, when you visit Family or a Friend for Christmas, Thanksgiving, Mardi Gras, Memorial Day or the 4th of July; after you Say Hello to everyone you Immediately ask for the Wi-Fi password..smh. A mounting body of global experts and research doctors suggests that extreme use of the internet is harmful to both physical and mental health. Why is the suicide rate in America higher especially in rural counties? After a decade-long study of the effects of screen time on more than 11,000 American children by the National Institutes of Health, those who spent more than two hours each day on screens scored lower on thinking and language tests. Please help me Lord B4 I go insane, nobody liked my pic now I'm a Zombie on them Brains—less than 3 face-to-face conversations a day with a human being, **Ya think!** Well clutch my pearls; this new age Cyber Digital Cocaine and Heroin are having a negative impact worldwide on people; regardless of race, gender, age, religion, socio-economic backgrounds; who are spending long hours playing video games, on the internet and social media causing them to disregard their family and obligations becoming virtually trapped in the virtual world as well as almost everyone consuming hours of negative media daily because **Fear is the Mind Killer**. The Smart Revolution is upon us now. Despite the fact that cell phone utilization has made life more convenient for many people globally, it has not come without some negative and adverse effects on the psychological and physical well-being and interpersonal relationships with humans. For example; because of the relatively easy access to the internet online community through smartphones, negative consequences as a result of a lack of restraint perpetuates impulsivity, and poor risk assessment in which the behavior becomes more widespread mainly in the forms of Cyber Bullying and Violence. Is it better to know some of the questions than having all the answers? How many bad decisions will you keep making before you die? How come some people with brilliant minds will see a fire and throw gasoline on it with their divisive rhetoric? These are just a few of the things I will address in this book B4 critical and independent thinking is banned. So, sit back n relax your behind bcuz I'm about to let loose and let u know what's on my mind. But before I do that let me ask you 2 dumb questions. Does whomever control the digital information and narratives, control and/or own you? Is marriage no longer viewed as a strategic partnership to create a legacy and generational wealth? Now that I have your antennas up—it's 1 for the money, 2 for the Show, Trumps in the White House, so eff it Let's Roll! Omg! Now where did I put my digital driver's license…..

Internet Ganstas, FB Bullies, Twitter Thugs, IG Fiends

SURVEY

1) Would U go on your Dream Vacation if your phone can't get a Signal? **2)** Is your phone like your Lover: it's the last thing you hold at night and the 1st thing U wake up to Every Morning? **3)** Would U parole a Social Media Cellphone inmate? **4)** Do U lay awake at Nite thinking about what's happening on Social Media? **5)** Do U wake up in the middle of the Nite to check stuff on Social Media? **6)** Do U send a group text msg saying hi just to get people to text you back? **7)** Is your to do list: Go on Social media, post, text, Eat, Selfie, play video games, watch youtube, eat again, Stream Live, Sleep, Repeat? **8)** Do U have a Fear of missing out (**FOMO**) so you Hold your phone in your hand with earbuds on while Eating? **9)** Do U get an anxiety or irrational fear of being without your phone? **10)** B4 you check on your spouse, kids or pet as soon as U wake up, do U check social media, text messages and emails? **11)** Do U prefer to Spend time on Social Media rather than having some Real face-to-face time with your Family and Friends? **12)** Do U Constantly look at your phone in meetings, church, while driving, at a party or out at dinner with family and friends? **13)** Do U lie to family n real friends about the amount of time u spend online? **14)** Has your Work productivity, quality family time or study decreased due to more time Spent on Social Media? **15)**

Would U rather get robbed or give up your purse/wallet than your phone? **16)** Would U say to a complete stranger excuse me: can I use your hotspot? **17)** Would U unfriend or block someone if they don't like your photo or post? **18)** Do U think it's ok for President Trump to clap back on Twitter? **19)** What is more addictive than nicotine, turning queens n2 selfie fiends? **20)** Does every generation make similar mistakes at each life's stage? Answering **Yes** to **5** or more of these questions means U have an Electronic Digital Dependency and not Brain Damage. If U refuse to answer any question after number 5 bcuz there's no app to help U means U have **the Internet Plague**.

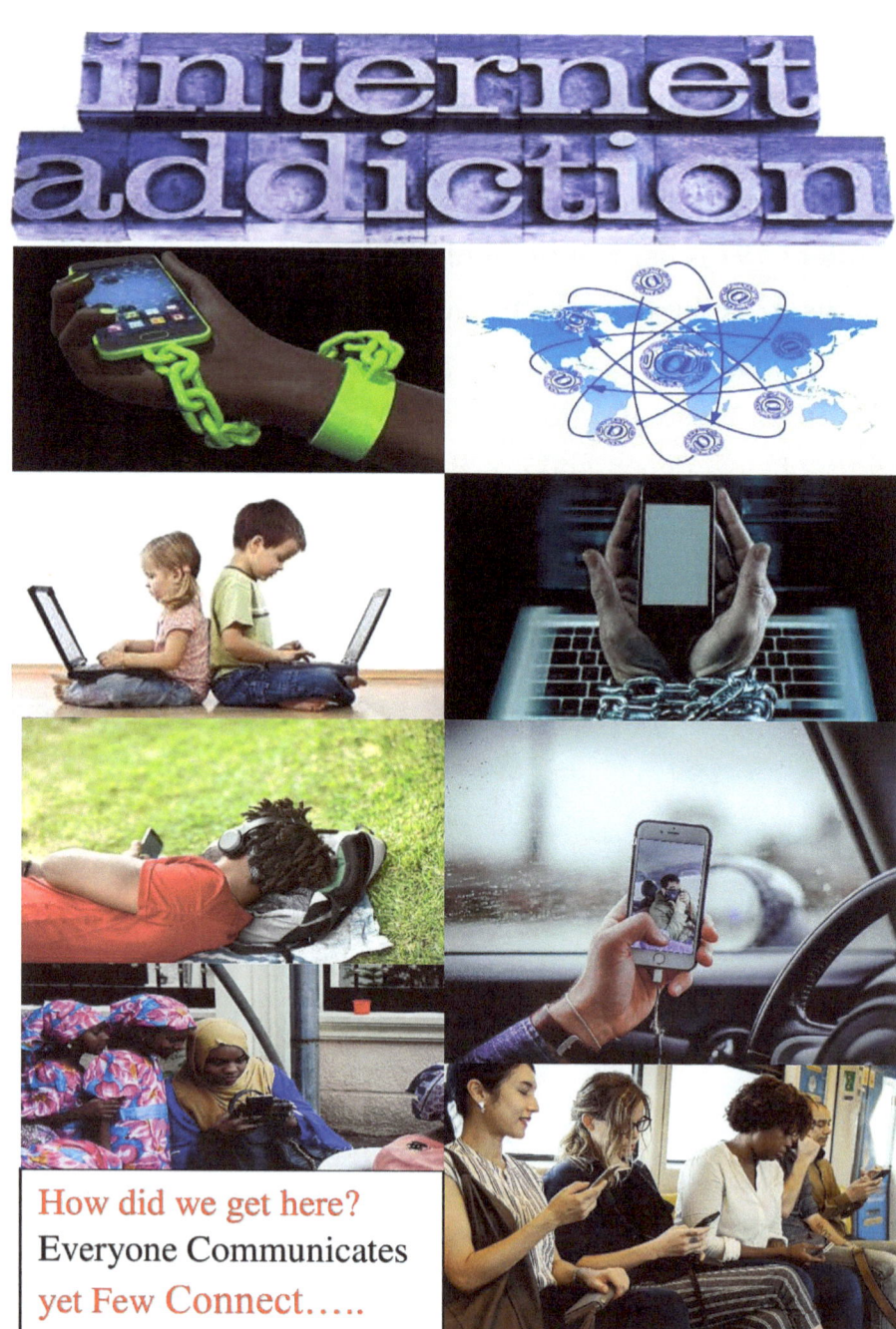

How did we get here?
Everyone Communicates yet Few Connect…..

Internet Dope Is My Life

To all U Internet Gangstas, Earbud Warriors, Bout that Street Life Wannabees, Facebook n kik Bullies, Twitter Thugs, Internet Trolls, Selfie Kings n Queens, Photo Bomb Kings n Queens n Instagram, Roblox, Snapchat n Fortnite Fiends—**the Internet Plague** is upon us and it's **Code name** is **Operation Self Hate n Terminate**. To all the Internet Dope Fiends and Shade Queens out there it means Information Brain Overload and the Countdown 2 Self Terminate Chatbot ai Revolution Style bcuz Vanity is your Insanity has begun. Phones down, heads up! Put your big boy N girl pants on, it's time to grow up. Why don't Americans over the age of 13 in the United States have any Digital Privacy, Legal Rights, Ownership or Control over the information that's being collected by Amazon, Facebook, Twitter and Google via voluntary mandatory permission or stealth consent coercion? How do U feel when someone is constantly looking over your shoulder in the Real World? If the answer is a resounding I don't like that; then why are so many people willing to be exploited knowingly—each time they use the internet they are putting themselves and privacy at risk because their ISP knows every website they visit as well as strangers monitoring their online activity and hackers capturing their financial transactions. Did I forget to mention all of those

Data Mining centers in Utah that eavesdrops on every American and is invisible from the highway surrounded by concrete walls, a security boom checkpoint with guards, sniffer dogs and cameras—not enough time to talk about that. Does America need a Spiritual Atom to touch down? Why do I feel good when I get a text or a tweet? Is social networking turning America's youth into Cyber and Digital substance abusers'? Are we supposed to declare a war on Cyber-drugs or a Cyberwar on Cyber-drugs? Are violent video games turning our males into trained assassins? Would you rather shake hands with the Devil or be in his path? How do you parole a Social Media Cellphone inmate? Now ask yourself, why do you park in a driveway and drive on a parkway. Money didn't make me happy—but why! How do you make Chaos your strength and profit from it? Is there any truth to this old statement: Hatred will destroy the container its carried in—while you watch your enemy go down it destroys you on the inside as well? Negativity and comparing yourself to others is the thief that steals our happiness—but why! Where is your emotional intelligence meter? What was the secret to the Hanging Gardens of Babylon? What happens to the mind when a person goes to solitary confinement? Where is your bitterness, mean spiritedness and hatred meter? Did feminist kill Chivalry? How are you conditioning your children's mind? What happened to all of the free newspapers? Would you send a soldier to Afghanistan with a bb gun? If not then why would you expose children to highly addictive devices before they have developed the social skills to deal with the complexities of emotional intelligence? I'll bet you a dollar to a donut that you won't find one politician on the Hill that will go on the record and address the issue of the long-term impact of technology on our future generations—simply because it can't be used as a political football and continue the divisive American rhetoric leaving our precious youth exposed to a clear and present danger. Where is your real life's happy meter? Does the Queen need to implement an individual mental aptitude improvement Brainasium weekly segment to improve the next generation's mindset? Shouldn't hemorrhoids be called Asteroids? Are we under Planetary Acquisition? Will ai unfriend and Block humans then Delete us? Can you sue a robot for causing an accident? What is more addictive than nicotine, turning future queens n2 selfie fiends and a twerk team? Will chatbots overtime become hateful, offensive and racist? What is the difference between Snitching, dry snitching and being a witness? Have we reached the tipping point where a generation of kids are afraid to go to school? With half of the American people High, some low; the other ones on social media feeling broke and don't know which way to go. How many different ways you can get fired from your job now since the Internet? Why is Facebook collecting data on people who don't even have a Facebook account? Why does the founder of Facebook Mark Zuckerberg want his own kids off of Facebook? Why does the average employee spend nearly 56 minutes per day using their cell phone and 42 minutes per day attending to personal tasks, running errands at work for non-work-related activities wasting nearly 8 full hours per week? What would your mental and visual retention logical bandwidth be like if there was no social media and smartphones? Who loses in a Gender war? Have you become Numb and just Don't Care about anything or anyone Anymore? Is attention or likes your new addiction and you'll do anything to get it? Is the Secret to Life Positivity? Hella screenshots and selfies til I die: "I text all day, I post all day: Is there any hope to get me off this Internet Dope!"—I wonder if he who knows 10,000 things (**Wan Shi Tong**) can answer these questions. There's only one right question for me: When will you believe what you're hearing with your own ears and seeing with thine own eyes to be a Clear and Present Danger to us all. There's a War going on for Control of your Mind. The Internet has been weaponized by those who want to spread Fear not just in America but worldwide perpetuating Evil deeds, bitterness, selfishness, greed, mean spiritedness and in some cases **Good ole fashioned Hate**. There's an old saying: the greatest sign of power is never having to use it—for all the millennials, Generations X,Y,Z and Alpha out there, that means U don't hv2 Flex or just imagine one of them dam Imma show this fool who they messing with emojis! Well now it's time we seek answers for some of the **$17 Trillion** pink elephant questions right now that everyone's been avoiding. Why isn't self-aware Tech and those hyper-aggressive **Digital Pimps** out West who covet and profit from personal identifiable information like **Facebook**, **Twitter**, **Amazon** and **Google** governed by the **Center for Digital Democracy** and the **Federal Government?** What type of **Social Media Citizen** are you? I can't believe I'm gonna utter these words: Intelligent and educated adults, mothers, fathers, sons and daughters seemingly cannot even resist the temptation to hold and check their phone even when they walk or drive down the street or drive on the expressway traveling at a high rate of speed texting, not even realizing the danger their putting their loved ones and other citizens lives in and well-being at great risk. smh! Is it time for local, state and federal government to legislate Stoopidity? Why are there no public health warnings on the boxes of Smartphones and Tablets like cigarettes? What is a Human Friend? If you get rid of anger, will that open up your 3rd eye chakra? Would you rather be Free or Rich? How much does it really cost to make a penny? Does a sneeze travel 100k mph? What is a follower? What does TMZ stand for? Are people really hungry for unfiltered information? I got 400 friends and not one of them know me or know the color of my bathroom at my house—some friend. Is Twitter and Social Media really what's destroying a future generation in America? Each year, over 330,000 accidents are caused by texting while driving lead to severe injuries. Each day in the U.S. approximately 9 people are killed and more than 1,000 injured in crashes that are reported involve a distracted driver. Why are there no age restrictions on **Smartphones** and **Social Media** like alcohol and gambling? Why are **Parents** still exposing their kids to highly addictive **Digital Drugs** or shall we say **Cyber Cocaine** and **Heroin;** before they have developed the social and **interpersonal skills** and the **coping mechanisms** to deal with stress or depression; and mastered any of the **natural defenses** of **emotional intelligence** to handle all of this access to instantaneous information and gratification causing **Internet** and **Smartphone Addiction?** If the eyes are the windows to the Soul, why are our young males afraid to make eye contact when introducing themselves or engaging in a conversation and females are not. I've witnessed this 1st hand in a plethora of public, social and business settings in which they don't give a firm handshake and won't look you in the eye while shaking your hand or holding a conversation. What are the alternatives for an undeveloped brain to deal with grown folk challenges, concerns and split-second life altering decisions in some instances? How do you have a healthy relationship with smartphones and social media in the virtual | digital world and form meaningful relationships with humans in the real world?

Our young people are no longer being creative or competing on a high level for some reason. Maybe because the last 2 generations, Millennials and generation Z were raised to think they were very, very special; and that they could become anything they dreamed of, and then after graduating from high school or college they found that the children of the Greatest Generation—the Baby Boomers; had let millions of jobs slip out of the country—did I say that out loud…my bad Democrats and Republicans (left-right-liberal-conservative-GOP-progressive-log cabin); I didn't mean to shine the spotlight on your duopoly you've been running for 150+ years and the people who voted you in has ended up getting that 80's slang term "the gas face." Yep I said it! Get over it! Just look at the impact technology has had on just one generation on Romance and the entertainment creativity of humans. Why are young men and women on college campuses making consent videos while dating to prevent them from being sued in the future just in case things get hot n steamy an someone has an agenda with ulterior motives? Is the Women's Rights Movement and the constantly changing rules of engagement between the sexes for Romance taking all the fun out of life? Have men really changed in over 1,000 years—food, sex, money, power and silence and not necessarily in that order? How do we teach the next generation to Learn to Look up Again? Is Smart Tech making the next Generation Dumber? In the last 20 years or so give or take a few, generation Z have only managed to create 24 new dances with little to no excuse: Choppa style, the leg kick, the keke, the Cha Cha Slide, Lean Back, the Jerk, the Nae Nae, the Stankey Leg, the Shimmy, the 1-2 Step, Walk it Out, the Dougie, the Soulja Boy, Twerking, Wobble Wobble, Gangnam Style, Line Dancing, the Dab, Chicken Noodle Soup, Cupid Shuffle, Halle Berry, Achy Breaky Heart, Lean Wit It Rock Wit It, the Harlem Shake, Dan rue Nick Nack Patty Whack. During that same time span of about 20 years from the mid 1970's to mid-1990's, 49 dances were created mind you with no internet or Youtube: Donkey Kong, Pee Wee Herman, Cabbage Patch, the Prep, the Bump, the Bus Stop, the Worm, the Bird, Electric Slide, Break Dancing, Running Man, Butterfly, Ed Lover, the Transformer, the Funky Chicken, the Snake, the Biz, the Robo Cop, push it, the Hustle, the Creep Dog, the Moon Walk, the Kid N Play, the Roger Rabbit, the Alf, YMCA, the Pendulum, Throw da D, the Eddie Bow, the Wop, the Reebok, the Smurf, the Troop, the Tootsie Roll, the Butterfly, the Macarena, the Bart Simpson, Da Butt, the Sprinkler, the Carlton, the Spank, Scrub Da Ground, Ride That Train, Air band, the Hammer, Twerking, Back Dat Thang Up, the Vogue, the Slow Drag and the Bougaloo—And yawl young people say we Lame! Ask yourself, why are our youth and some adults turning to social media and a device to deal with stress, depression and rejection instead of turning to someone who Loves them like a family member or friend subsequently becoming a "mobile drug addict" resulting in that person feeling like they cannot function without it in their world i.e. FOMO. In my opinion, this contributes to more anti-social behavior for some causing an addiction to their device creating: Social Media Zombies, Internet Gangstas, Earbud Warriors, Instagram Queens, Twitter Thugs, Selfie Soldiers, Snapchat Fiends and Facebook Bullies! Bullies back in the 80's would jack you up at school, take your lunch money, take your chocolate milk, BBQ corn chips and funyuns, slap you on the back of the head, give you a wedgie, knock your books out of your hand in the hallway and everybody would be laughing at you. For a little fun, rhyme this True story: "U 5 foot 10 and I'm 5 foot 2; now that's 200lbs against 132. Bully just mushed me so what I'm gone do. How did this story play out back in 82? Pull up in the coup de ville to the school with all my kin! Pointed out the bully and now the saga begins. So funny how the school thug needed a hug, cause of the hole he dug when he saw my cuz, who's 6 foot 2. Now the bully has bubble guts and has to boo boo too. Now he's running thru the school calling for help, screaming people trying to hurt me n they everywhere. Now the bully's on the run like Elmer Fudd, chasing Bugs Bunny wit his gun. And now the tables are turned, on Elmer Fudd, Ain't no fun when the wabbit has the gun. Gave the bully; a 2-piece fist sandwich now his eyes lips n nose covered with bandages"—Bully problem solved! It was hard to be a Bully with a Bloody nose, black eye, busted lip and sore jaw back then. What will your kids be saying 20 years from now? 1. My number one priority is figuring out how to keep a roof over my head and food on the table. 2. My basic needs are covered, but my debt is overwhelming—paying down my debt is my number one priority. 3. My financial picture is stable as long as I am pennywise n not dollar stoopid—my main priority is spending wisely and saving for the future. 4. My financial picture is stable, but I want more! 5. My main priority is learning what I need to learn in order to move from safe and stable to spectacular and magnificent. 6. I am fine financially; I'm really fortunate when I think about it, but the truth is I feel dissatisfied and my main priority is to discover what's missing in my life and do what I have to do to have a joyful and fulfilling life. 7. I'm an accomplished professional, I have a solid career, but I feel like I've reached a ceiling when it comes to my income and my main priority is to learn what I need to learn and do what I need to do to smash through to the other side and I'm not afraid to work hard to do it! America is now on the Cusp of having a generation of people who prefers technology over humans and often walk around connected virtually with universal do not disturb signs on called headphones and earbuds closed off from the real world, not to mention texting and driving which is a whole other discussion. Instead of living until they die, they are surviving until they die and that must be a scary place for them. It's not entirely all their fault since they just happen to grow up in an era where the 1962 futuristic cartoon, The Jetsons elaborate robotic contraptions is now part of everyday life almost 60 years later. Who would've thought that what was depicted in a cartoon six (6) decades ago including whimsical technological inventions which has now landed on our Youth doorstep and their family and sworn protectors purchase these expensive electronic devices for them; because they can simply track their physical movements creating the illusion that it's all done to keep them safe, connected and it's really, really, cool? The only thing missing is an escalator in their home and using an airplane to go to work daily because they have all the other savvy tech: Robots for a pet, face time or skype anyone they choose…etc., etc., etc. Unbeknownst to these limited thinking adults who often reward and justify bad behavior because they're addicted to the Digital Drug too, the absolute unhealthiest thing you can do is promote anti-social behavior with our youth instead of observing and taking their mental aptitude temperature to assess their emotional intelligence, life, social and interpersonal skills development. Am I Using Technology or Is it Using Me? If U don't know, U better ask somebody with real White or Grey hair. Don't bother asking some of these millennials because they're always looking down never up and most of them are Social Media Zombies, Internet Gangstas, Earbud

Warriors, Selfie Soldiers, Twitter Thugs, Internet Dope Fiends, Fortnite Junkie and on the verge of having the 1st digital baby called Lane Pee Brain. Just the other day on the Internet, a Dude was on his phone texting and walking with his universal do not disturb sign on; and fell N2 Tony the Tigers' Den! Tony said thru his Cage after taking a Selfie, Wow! **He Tasted Grreeaat!** N my post of me Eating him Just Got Hella Likes on FB n IG lol. Why does Facebook and your Smartphone have the ability to sense your mood in any room? Does the Medusa Lake of Africa really turn animals into stone? When was the last time you had an idea? Has America become anti-family? Who or what is Really keeping this Country and your Family together? Does a Strong family equal a Strong Country? Who or what is trying to Destroy your Family? Is it better to know some of the questions than having all the answers? What has happened to Critical, Independent or out-of-the-box Thinking with our Youth? How come a windshield is 100 times larger than a rear-view mirror? Why does the Government or Media give you a target or boogeyman of whom your perceived enemy is or who or what must be hated and then eliminated? Should we continue to Destroy everything as we once knew it and recreate it in thine own Image? Is it normal to feel like you can't Concentrate because your Attention Span is in a state of Fragmented Reality throughout the day? Why are some people traumatized when they are unfriended or Blocked on Social

Media? What has really happened to the mindset of our youth in America? Not so long ago I can remember that as a 5-year-old, that touching a hot stove was a Lesson Learned long in life. Nowadays our youth are willing to put their young innocent lives at risk trying to become the latest video sensation on Social Media and Youtube with Stooopid challenges that would have Max Headroom, Mother Goose, Peter Pan, Wylie Coyote and Foghorn Leghorn scratching their heads: like the Netflix walking around in the Mall with a blindfold on challenge, eating a tablespoon of cinnamon challenge, lighting yourself on fire in the shower, jumping out of a moving car doing the keke dance challenge, vaping synthetic marijuana, eating tide pods, doing the Kylie Jenner shot glass lip challenge, eating 150 pieces of warhead sour candy in 10 minutes, the salt and ice causing the skin to burn challenge, eating the most-hot peppers challenge, the eraser on skin challenge in 2012, throwing boiling hot water on your body challenge, the choking game challenge, neknominations drinking challenge, the eating whole garlic challenge, the water in a condom challenge from 2006, car surfing, game of 72 runaway challenge, drinking vodka thru your eyes challenge, the cactus eating challenge, the taser gun challenge, doing the most cocaine challenge on video, the planking challenge and the fainting challenge. What about a Walk Around with some Sense looking up Challenge for these Manure Chocolate Smoothie Silicon Valley Attention

Engineers or as they say down South, some Sugar on Shit Sweet Tea? Or better yet: How about some positive skill and determination Good Ole fashioned We're Better than you Neighborhood Olympic challenges that doesn't result in a permanent scarred body, brain damage or death: red light green light challenge, a double-dutch Jump rope challenge, Hop Scotch challenge, one-legged race challenge, see who's the fastest foot race challenge, egg toss challenge, Shaper super toe super jock football game challenge, water balloon toss challenge, tug of war challenge, boys vs girls relay race, boy and girl piggy back race challenge, human wheel barrel challenge, human cart wheel challenge, hula hoop challenge, frisbee throwing challenge, arm wrestling challenge, thumb wrestling challenge, put on them boxing gloves n settle that beef challenge, standing in the corner on one leg B4 U fall challenge, girl on boy shoulders foot race challenge, how many books can you hold b4 you drop them challenge, fast power walk race challenge, stare who blinks 1st loses challenge, tickle who laughs 1st challenge, Paper football game challenge, soul train dance move challenge, salsa dance challenge, group line dance challenge, spit ball in straw challenge, flying paper origami airplanes all the way from the back of the classroom hitting the teacher writing on the dry erase board challenge—well maybe not the last one because with nowadays teachers this would prolly be a Felony or proscribed conduct consistent with terroristic threats. Even the Slang in 2018 will leave you scratching your noggin. **Back in the 60's** people were saying slang terms like: it was a Gas, all show no go, beat feet, bench racing and bad ass. **In the 70's** it was skinney, can you dig it, bunny, brick house, boogie, psyche, bugged out, far out, get down, chill, fab, feel the funk, phony, you big dummy, going to the crib or gig, cheese-eater, jack squat, catch you on the flip side, awesome possum, jelly brain, jive turkey, Bam, Groovy baby, cool beans, the joint, Gully, the man, be there or be square, soul brother, keep it pimpin, peace out, mind your potatoes and awe sooky sooky now. **In the 80's** it was: Bad, fine, No duh, crunk, dag, fugly, lame, illin, rush, hacker, glam, fricking, the gas face, pac-man, chump, homeboys, one love, peace, no can do, bunk, chump change, weak stuff, duke it out, hip, take a chill pill, cool out, cold, Mickey D's, doofus, hair bear, totally legit, amped, punk, fat slob, posse, barf me out, like totally, yuppie, buppie, what's the 411, cowabunga, I kid you not, where's the beef, Bangin, Bite me, wig out, the bomb, keep it gully, yikes, totally awesome, druggies, I'm down, let's book, shiz nits, yello, bootleg, boss, let's bounce, the bombdigity, trippin, jam, bud, skeezer, shut up, schweet, slam dance, step off, butt ugly, sportin, scumbag, scaredy cat, What's Crackalackin' and sick. **In the 90's** it was: dawg, home skillet, deuces, hold up playboy, hey fat, off the heasy, homie, let's dip, bezzled out, baby daddy, dude, the GOAT, the bomb, fine, fly, dope, hey cuz, drive by, RIP, head busser, wet boy, toe tag'em, smoke that fool, break yo self-fool, run dat, Buzz kill, I'm Gucci, hyped, mofo, cheddar, dead presidents, cha-ching, all about the benjamins baby, flawsen, fool, Sup, oh snap, O.G., wangsta, hate the player not the game, no doubt, chill out, FYI, dubs, Beeotch, all that and a bag of chips, it's all good, wiggity whack, duh, homeslice, score, guap, bling bling, ice, phat, ghetto, mad good, chillin, jack you up, wife beater, booyah, yo yo homeboy, triflin, keep it real, player hatin, hoochie mama, ya damn skippy, tight, two snaps up, talk to the hand, open up a can of whoop ass, throw down, bootin up, quit icing my grill, scrub, my bad, don't go there, going postal, let's roll, po po, fresh, as if, freakin, gank, blazed, punani, o hell naw, jack, brothers from another mother, f-bomb, eat my shorts, what's the dillio, fart knocker, come holla at me, hard, get over it, shady, hella cool, getting jiggy, word, you straight, yadda yadda yadda, you be trippin fool and hottie. **In the early 2000's:** Bling, Baller, tricked out, Biotch, dime, Crunk, tipsy, turn up, lit, dawg, trashed, that's hot, boo, you got served or clap back, I own you hoe, aiight, stoked, true dat, peace out, bounce, dope, what's poppin, wassup, what's cracking, da bomb and wangster. **In 2018** the Ghetto Bots and Angry Thots slang terms are: swag, turn up, lit and **drip drip**. To all the young people out there under 30, for the last 50 years drip drip meant you have a STD and as Kool Moe Dee said, "Go see the Doctor." Now back to their 2018 slang communication: pull up, rip, tbt, bb, ilysm, lol, smh, omg, wtf, lmao, lmbo, tf, smol, mf, bc, ik, ikr, gd, fam, salty, thirsty, otp, throw shade, @me next time, bae, extra, ship, go ham, glow up, go off, sus, snatched, boots, sis or you mad bro, stan, thot, hunty, doggo, swole, oppo, clap back, throw shade, ratchet, trill, dm, woke, Bible, Gucci and **keep it 100**. Is this the type of America you want your **Children** and **Grand Children** to continually grow up in: Democrats, Republicans, Governors, Mayors, Judges, Sheriffs, CEO's, Chairman of the Boards, Superintendents, Principals, Teachers, College & University Presidents, Professors, State Legislators, Professional Sports Owners, Entertainers, Philanthropists, Wall Street, Hedge Funds, Retail Industry, Oil Industry, Tech Industry, Automotive Industry, Communications Industry, Agricultural Agrochemical Industry, Bio-Tech Industry, Food Industry, Pharmaceutical Industry, Clothing Industry, Consumer Goods Industry and Religious Institutions? We are never going to agree on **Race**, **Politics**, **Culture**, **Abortion**, **Religion** and **War**; and the things we consider **morally** acceptable or morally wrong: **Suicide**, **Cloning**, **LGBT**, **Marital Cheating**, **Divorce**, **Polygamy**, **sugar in grits**, **sweet tea**, **sugar in tuna fish** and **spaghetti sauce** or the **Death Penalty**. There's not enough time to peel back the onion on these mind-boggling conundrums-- my Twitter Finger is now a trigger finger. There is nothing wrong with Smartphones and Social Media but too much of it can cause anti-social behavior and depression. To all the certified smart people out there: What Does a Spider Use to Catch Human Prey? What's a maybe Junkie? Ya got more degrees than a thermometer but no diploma from the university of walk around sense. What are the side effects of too much Cyber Cocaine? What's more important than taking care of your brain? How do I deactivate my Hate? What type of lyrical and visual medicine do you feed your brain on a daily basis? Has humanity been conquered by technology? Will technology Kill a future generation? How do I become an Attention Engineer? Is Lady Justice Really Blind or does she just wear a blindfold when powerful and famous people commit criminal acts? What's the long-term effect of too much information? How do u see what u can't see? Has the internet made your brain a pea? I'm the enemy with no Face, who am I. I have over 400 friends and not one of them know me. Do our **Brains** really want us to nurture and preserve a **fortress** of **illusion**? Is there a Cure for your Delusion of Inclusion? What's on your tree of shame? How do you live with your distorted view of the uncomfortable truth? When you look in your Life's rear-view mirror, do you see how far you've come or is your life still stuck there—going back to a college homecoming 25yrs later just to compare your progress against someone whom there was a mutual dislike. Why am I so obsessed with pleasing people? Do you suffer from an electronic dependency or Smartphone Addiction? Does everyone deserve the right to have high self-esteem? Who are you and what do you want? What on earth am I here for? What kind of reputation do you want? What is Social Media Etiquette? How do you master emotional intelligence in a digital world? If Roseanne Barr had contacted me B4 her emotional response in this new digital world, I could've saved her millions of dollars and saved her reputation because an emotional or verbal response to a tweet, text, email or voice mail can have severe life altering consequences if you don't master emotional intelligence. Before the internet, you would have to go to the archives, microfilm or find an old newspaper article to dig up dirt on anyone over the age of 43 lol. In the Digital World almost everyone has a permanent transcript or video evidence of their transgressions, comments and emotional responses. How come critical, independent and out-of-the box thinking is absent from higher learning institutions in the free world? If U live in a box and think in a box then all of your decisions and choices have to come from within the box; but once you get out of the box then your choices are limit-less, and the sky is the limit. Would you rather be right and still be mad or would you rather be wrong and be happy? How do you recover from intellectual fatigue? Do you have Fomo? Why is Attention and Likes your new addiction and you'll do anything to get it? Can you get a PHD from Hustle University? Does anyone have the antidote to cure my Nomofobia? How come this generation will pull a trigger with no hesitation—that twitter finger is now a trigger finger? How many people really love and care about you? Why do some of us need validation from our sworn enemy? Why are the 7 Wonders of the World not the Moon, Sun, Stars, Clouds, Water, Fire and Wind? Why is Fear the mind killer? How do you Create positive outcomes from difficult conversations? Why does everyone communicate and yet few people connect? When Life punches you in the Gut, what do you do? And finally; to all my ThinQ4self **thinkaholics** out there, why do people from my hood spend beaucoup money they don't have, buying stuff they can't afford, trying to impress people they busy hating on everyday all day! Fast money is slow time Playa n Playettes. Git your game out your pocket trust fund babies, athletes, entertainers and hot boys! And in the Suburbs, just give that kid a **Cookie** and an **iPhone** Mommy—that'll keep 'em quiet until the battery dies… **Kewl Parenting 101**…lol! Well clutch the pearls and shut the front door; but you just opened up the back door to that beautiful Macedonian phalanx for little Dante, Skyler, Santiago, Amber, Julio, Anika, Lee, Sara, Brandon, Mei, Muhammad, Maria, Sofia, Ashley, Felipe and Laquan. Why does America allow its Youth to isolate themselves? Nobody is Really who u think they are. Americans over the age of 13 in the United States currently does not have any Digital Privacy, Legal Rights, Ownership or Control over the information that's being collected by Amazon,

Facebook, Twitter and Google. These digital pimps, attention engineers and modern-day hidden dragons have low jacked a generation and a future generation of humanity's brains in the free world with military like precision like a thief in the night and here's why. Whether you own an Android or iPhone, you still need to give consent in order for the firmware and software to fully integrate and use many of the device apps like Facebook, Google, Twitter, Amazon, Instagram, Banking and so on. This voluntary mandatory permission or stealth consent coercion is diabolically clever and highly profitable—can't wait to see what the master plan is for the highly sought after thousand dollar plus new hologram and infinity flex phones in 2019. That's just what our future generation needs, more screen time and voluntary solitary confinement creating more time bomb ticking assassins headed off to school to use lethal force on their classmates. Everyone is going thru something in their Real Life. We all have a Shadow Inside Us. Everyone puts that Mask on to Cover up all their weaknesses and Sadness; and even with all of that intelligence—be it street smarts or book smarts they've Built up these Walls to keep people out not Realizing that they're the one trapped inside leaving them all alone. Spending 70-80% of your life in survival mode envisioning the worst-case Scenario not even enjoying life anymore by not living; keeping your Mind and Body in a State of Fear is the Uncomfortable Unpleasant Truth we don't want to stomach. What costs are you willing to pay for Vanity? We live in an age now where a new sub-human species of gender promotes abnormal earth, but what appears to be normal self-promoting inter-galactic display of rainbow hair colors not just in America but Worldwide. Thru all of my research, there are only 5 natural hair colors in the world: Black, Brown, Natural Red, Blonde and White/Grey—Just ask the Government and Clairol if you don't believe me. There are no documented records of any human species being born on Earth today or in the past with rainbow, lucky charms and a peacock like hair color. Do you know how many people have died at the hands of the Doctors of Death from altering their body thru vanity enhancements? There's a whole generation in America that has never had a bloody nose, bloody lip or a black eye or even been in a real fight where punches are thrown for that matter but is willing to kill in a New York second—almost like a Gladiator mentality: either there's victory or there's death by lethal force. Some call it bullying versus those half-truths or bold face Comforting Lies we would rather tell and hear that creates Delusions of Grandeur in the Virtual World turning future wives and queens into virtual selfie fiends and twerking machines. Life is Unpredictable and short but More Unpredictable than Short. When you make all those withdrawals from your emotional bank account without making any positive deposits; it leaves your Happy Meter bank account out of Balance with

Insufficient Funds putting your life's compass on collision course with emotional bankruptcy. My goal is not to reach every person but the right person because negativity and comparing yourself to others is the thief that steals your happiness. Let's take that mask off and meet the real you! Let's stop ignoring those pink elephants in the room and create an environment to talk about the difficult things that divide us and use our Emotional Intelligence to create positive outcomes. We are not going to be able to donate and go fund me our way out of this quagmire America finds itself in or should I dare say Clear and Present Danger or the Crouching Tiger hidden in plain sight—How do U see what U can't see?....even a 3rd grader knows this? I want to enrich the Lives of future generations and help them Build on what they wanna build on to make a positive impactful Trigger in their universal thought process shifting their Mindset by speaking their language-- Listening to Learn Not to Medicate, Judge and Earn! Millennials and young people still eat, sleep, cry, breathe and spend money; so, there's nothing fundamentally wrong with these young and grown up kids even though their entire life is in their phone. In public they often stare into their phones never looking up even if they're in the company of another human and is seemingly terrified if they have to make eye contact and engage in verbal conversation and don't have a clue about their nonverbal intelligence which is a critical skill that is needed to thrive in the real world. If u believe in peace, let's keep it. If you keep an open mind, you'll be pleasantly surprised what you can accomplish. Accept who you are. When you look in the mirror you should see the image of God—You should not see any flaws, you are altogether beautiful. I see no flaw in you. You are fearfully and wonderfully made. There is no flaw in you. Everything you need is already inside you. Beauty is in the eye of the beholder anyway. Tell me who you are! Humans weren't created to have a hive mentality like bees and ants; Act the same and look the same. We are unique. Not even identical twins have the same fingerprint. What separates you from the competition? Furthermore; 77% of adults and 55% of teenage drivers say that they can easily manage texting while driving and 48% of kids in their younger teenage years have been in a car while the driver was texting; 35% of 18-25 year old's said in a study that they can't eat a meal without holding or looking at their phone so Intellectualism coming from a hypocrite creates a double standard societal conundrum. Before Columbine, the mindset for kids in middle to high school were either going on to College, the military, finding a job, working in the family business, going to the pros, dating boys or girls or starting a Family and not worrying about one of their classmates coming to school with an AK47, AR15, shotgun, Glock or 9MM to Shoot as many people as possible. More weapons, more security and more laws will not solve this problem with the Youth in America. After several conversations with our Youth from all over the country; LA, Chicago, Atlanta, New York, New Orleans, Birmingham, Detroit, Miami, Phoenix, DC, Baltimore, Dallas, Houston, Memphis and St. Louis the overwhelming response was that "No Disrespect Man! We don't care how much you know until we know how much you care! Nobody loves us anyway, so we turn up every day…don't really know why I text and drive but I'm ready to die. The Government goes and protect people in other countries and show us no love and don't protect us in our own country. Our so called, Politicians and Leaders have Failed Us, so what can we do at this Point." The only force capable of Transforming this generations mindset is Love and Peace. Hatred and Jealousy brings War and Murder! Love and Peace Brings Harmony and Joy! The jails and prisons are Full in America, China, Spain, Italy, France, Ireland, Australia, England, Canada, Central & South America plus several more with our Young people. By 2025 Schools will be like a Supermax Prison Facility with armed Resource Guards, Cyborgs and Robots on Patrol outfitted with the best X-ray equipment including Retina Scans and Thumb print scanners just to get into School plus all the Robot Teachers packing HEAT and a FAA Compliant HD Cinematography drone assigned to each teacher and principal. Our Youth are Crying out for Help and No One is Listening to the way they Communicate. They Communicate much differently than our Youth did 25-30 years ago, and they don't even realize the potential consequences and the pitfalls of Social Media, FOMO, Nomophobia, Smartphone Addiction and having a permanent transcript of everything they say and do and how it can be made public at any time. There is nothing wrong with having a healthy relationship with Technology but when it becomes an Addiction then that's an issue that must be addressed without infringing on our Constitutional Rights. However, some experts will argue the facts; some experts will argue the law; and if neither applies they will cause confusion engaging in Cloak and Dagger techniques that is hard to make by the average adult let alone our youth. Is this the type of Learning and Lifelong Experience you want for our youth, your future Grand Kids, Cousins, Nieces and Nephews? Instead of strategizing to navigates one's Life compass for Success, our youth are strategizing like they're going to War on their school classmates and anyone in their Path—causing significant Collateral Damage across all Demographics. Life is Unpredictable and Short but More Unpredictable than Short. Still today, the mantra is still the same: If there is no Struggle there is no Progress. I bet Dr. King, Thurgood Marshall, J.L. Chestnut, Jr., Presidents Clinton, Bush, Trump, Obama and Reagan didn't envision this Madness. America is on the Cusp of a gender war, race war, political war & All OUT Nuclear War... Are we in the last days? Are we killing each other too fast? The Culling of the Herd (the innocent) has

Begun and we haven't even explored the Dynamics of Racial Tolerance, improving communications across racial/ethnic groups, how to identify the 3 levels of racism today and create positive outcomes from difficult conversations. Despite our perceived differences, the moment we Stop fighting for each other is the moment we Lose Our Humanity. Is it time for Every U.S. Senator, U.S. Congress Person, State Senator and Congress Person, Governor and Mayor in America do what's Unpopular—their Job! Protect us from all Foreign and Domestic Threats. Have the American people lost all confidence in a Government that appears to be broken with no legitimacy, resorting to finger pointing, he say she say drama pitting mostly limited group thinking divided fearful citizens choosing between diarrhea and constipation in the voting booth and issues no one will agree on? It's the perfect utopian hustle recipe and hunting ground for savvy Professional Career politicians rendering legal citizens cries for help futile like a wildebeest in the jaws of a 20-foot Nile crocodile. Furthermore, you can't solve a gun and bullet problem with more guns and bullets. There is a way to get our young people back on track. After over 222 years of being at war and fighting with each other; subsequently another 45 years of the war on crime and 47 years of the war on drugs; it has left America torn, hurt and so divided you can cut the tension with a piece of wonder bread burnt toast. But with more love, respect, peace and understanding we can put it back on the right path. When change comes—you can either participate or react to it. Why is it lonely at the top? Is life easier when you fail, and you have to look into that mirror because you know you have to get better? Why is Everything out of Balance? There are 7 Energy Centers in Our Body in which Energy Flows thru. The Word 'chakra' means 'Wheel of Spinning Energy'. A chakra is like a Whirling, Vortex-like, Powerhouse of Positive Energy that is lacking in America. Within every failure lies the seeds of Success. Close your eyes, count backwards slowly: 3-2-1, then open your eyes and say this out loud to yourself 3 times: **"Today is the day I get rid of my emotional baggage and bury it bcuz I keep making all these withdrawals from my emotional bank account without making any positive deposits leaving my heart heavy with insufficient funds with no overdraft protection. The mind governed by the flesh is death. The mind governed by the spirit is life and peace."** Everyone deserves the right to have high self-esteem. Your talent will bring you them riches, but education will help you keep and grow those same riches. Education is education, whether you work at a **Hedge Fund** with **$3,000 John Lobb's** on your Feet, or you're serving your people out the **Trap** with **$300 Jay's** on your Feet—**Fast Money** is Slow time playa no matter how you slice it up. Just like there's a playbook to execute the game plan in football, there's a playbook to life and business as well or maybe it's better to have some good luck. Some will define Good Luck as when opportunity and preparation meets; but when things don't go your way you only have 2 choices: either curl up into a fetal position and call mommy or you can continue to work hard and get better. Do you want to win or do you want to be the reason you win? Every Human has the capability of either being the hammer or the nail whether it is physical or emotional and it is not gender-neutral truth be told. Study your enemy B4 confronting them! What you know is dangerous 2 your enemy; what you think you know is dangerous 2u. Sometimes words can hit hard as a Mike Tyson fist! Real Fake vs On the Real: fake life, fake wife, Fake eyes, fake thighs, fake butt, fake gut, fake breast no rest, fake purse, don't make me curse, fake hair—I don't care until I get fake money. Well call the **FBI** for crying out loud! Somebody just gave me a Fake **$20** bill…well consider it a tip to go towards all the real **$$$** you already spent on your **fake life**! Thou shalt know the Truth and the Truth shalt make you Mad—**Selfie!** ☺ Remember, the same mistake made more than once is a decision… Don't chase money if you're not an entrepreneur, find a stable company with awesome coworkers and a great boss because there's a ton of money out there but few great places to work with great bosses. Don't take anyone into your tomorrow who doesn't support you today! The winds of change are upon us. When the elite can't say what they want, nobody will be able to say anything without persecution or scrutiny in America.

Problems created by Man Can be Solved by Man and it's Better to know some of the questions than having all the answers. Am I Dead Wrong or Am I on Point? What Say You America? Here's a quick rhyme you can rap to your favorite 90's rap instrumental by Biggie, 2pac, Dr. Dre or BG: **"Insurance on my phone but none on my Life; my skinny jeans too tight but that's aight; stayed up all night and I can't even fight; Internet Dope—Is my Life; Do something with your life and use your Brain, after a while on the Internet it's all the same: Post another pic; send another tweet; bout to hit dem streets with these Jays on my feet. Hella screenshots and selfies til I die, look at my timeline it'll make U cry; I got 400 friends and not one of them know me; 75 likes—I told ya! I text all day, I post all day: Is there any hope to get me off this Internet Dope!"**—Request your Facebook, Twitter and Amazon Dossier today B4 its 2 Late—It's Free!!!! The Future Waits for No One. Love is when the other person's happiness is more important than your own, so get rid of that fake hair color, fake hair, fake eye lashes, fake persona and take that mask off so the World can meet the Real U. Stop making checker moves in Life's game of Chess—So much for Turbo Charging Creativity and Innovation. Can't wait to see Breaking News Coverage on one of the 3 letter networks of the: Gucci Ophidia Clutch power charging wallet for those smartphones, the 1st marriage of a Human to a Chatbot and Americans texting while streaming live and driving on those new flying motorcycles and cars in 2020. **#youBigDummy**. When anyone says to be perfectly honest to you—you best believe the next sentence out of their mouth will be a Real thick Organic Manure Milkshake—that's Cow Patte' Boisson if you're French—why lie when the truth will do. If U still need answers to some of the questions, go talk to someone with real gray or white hair who Loves U or just hmu @tiagllciad on FB n IG or taboris@tiagllciad.com. If I offended U, take a cry break; count backwards 3-2-1 out loud with your eyes closed then look into a real mirror and not your phone and say this to that unique human u see: who are U and what do U want—smile then Ghetto 1st lady wave to all of that negativity, low self-esteem, selfishness, mean spiritedness, ratchetness, mean mugging, drama, bigotry, self-hatred, skin color prejudice and living pay check to pay check broke b4 payday aka ghetto fabulous. No need to further engage in hyperbole….go back to your fake life bcuz all social media does is allow unprofessional people disguised as actors with no real talent an arena and stage to perform i.e. a Solopreneur! Lol…..If your life is boring and crappy, do something kind and positive and make it happy. And be glad you have a dad, so don't get mad or feel sad if you're a lad named Brad or Chad wearing plaid at your pad; and writing down your world view on that notepad is just a tad bit rad and facetime me anytime on an iPad even if you live in

Cuidad." Does everybody cheat in some capacity? What happened to the village discipline for kids in which parents and neighbors used a system that worked for centuries for many civilizations? Well, we can debate that forever on a blog or radio call-in show; but is it finally time for us to question everything since there are more open-minded people in the free world. I can remember a time when friends would take turns pushing each other on the swing. Then no matter what the weight difference was between them, they would get on the see saw and make the best of it sometimes defying the laws physics. When will we realize that no app, electronic device, post, tweet, text, email, direct mail, website, politician, reality tv show, billboard or video is needed for Love? It's finally time for everyone to stop being so close minded. And look within to discover the roadblocks and boundaries Life has thrown at you to help you unlock your true potential and stop the blame game and finger pointing. Since critical and out of the box thinking has been eradicated in the western world, if you lost your eyes where would you look first. Are we lost or are we in a state of fragmented Reality with no Soul? Should any human from the age of 15-75 be trusted who text and drive safely?

What is a Maybe junkie?

A maybe Junkie

is a person who

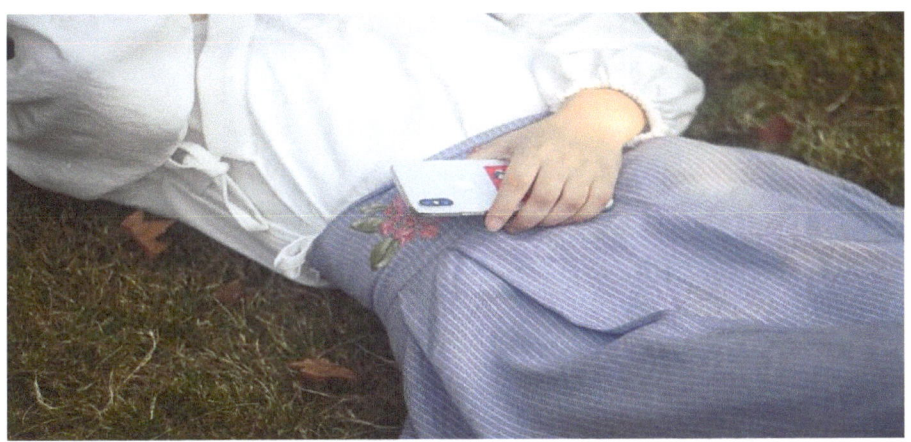

says to themselves:

Let me check my phone

every 5 minutes; maybe I got a text,

tweet, instant message or a like,

tagged in a photo or

somebody changed their

profile pic because their phone

is not in their life—their life is in their phone

sbt

Will Technology Kill A Future Generation?

Am I using Technology?

Or is it Using Me?

What does a Spider

use 2

Catch

Human Prey?

Everyone Deserves the Right to have High Self Esteem

Within Every Failure Lies the Seeds of Success

The Eyes are the Windows to the Soul!

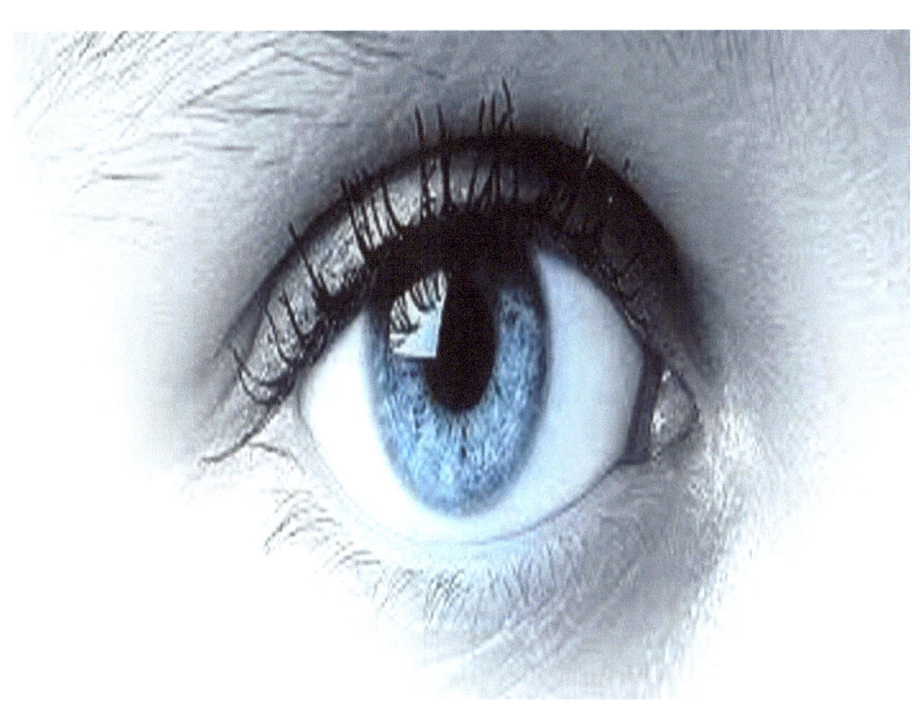

The Secret

2 Life

is

Positivity!!!

Negativity

&

Comparing

yourself to others

is the

Thief

that

Steals your Happiness!

Is it Finally time,

for us 2?

Everything?

A Good

Leader

must use

Wisdom,

Emotional Intelligence

2 for See how

Terrible Decisions

lead to Regretful

Results!

When there's an Earthquake

in America,

whose Fault is it then News Media?

Will texting, posting, getting Mad or Tweeting about Something

Solve All of My Problems?

Why do they wanna keep us depressed, ignorant, fighting with each other, angry, greedy, jealous, idolizing vanity, hating our lives and family with the help of technology? And just maybe and I mean just maybe: Ms. Alanis Obomsawin was right when she said; "When the last tree is cut, the last fish is caught, and the last river is polluted; when to breathe the air is sickening, you will realize, too late, that wealth is not in bank accounts and that you can't eat money."

Internet Plague Antidote:

1st Hit the Real-World Mind Reboot button by taking a deep breath n counting backwards 3-2-1 with your eyes closed; 2nd sit down face-to-face with someone who Loves U and ask them verbally to tell U 5 things they like and don't like about U and record it with ur phone in airplane mode n say thank u when the trama ends. 3rd Process it n Believe it. 4th sit down face-to-face with another person who Loves U and ask them verbally to tell U 5 things they like and don't like about U and record it with ur phone in airplane mode, say thank u and evaluate it. 5th sit down face-to-face with someone who U think Hates U and ask them verbally to tell U 3 things they don't like about U n say thank u for helping me and being my real friend. 6th Knowing is half the battle; now Stop being a Robot n Live ur Real Life and invite those 3 people out to lunch or dinner for keeping it 100 with u face-to-face—telling u what u needed to hear vs what u wanted to hear. If I offended or shocked U with my honesty and candor—git over it; bcuz now I have U doing what's unpopular in America—challenge U and made U ThinQ4self,..Now U don't hv2 pull n2 the drive thru an order one of those 1-16 value meal combos anymore. Here's a New Slogan 4 ya news media: Make America ThinQ Again! Perhaps the Navel of Humanity needs to be pressure washed since it's the most important point in the Human Body. Furthermore; it is very unsettling that on a daily basis how do these smart, intelligent 6 figure earning adults tell young people to put down their electronic devices when they themselves walk around in their presence in their homes, with expensive earbuds on listening to e-books n podcasts laughing out loud, ignoring the hell out of them and will scream and yell at them if they dare ask them a very simple question or want to talk about what happened throughout their day—now that's more Kewl parenting 101 ya damn Digital Bigot. Try this iPhone/Android and a Cookie Kewl Parenting 101 exercise for 3–9year old. Put the electronic device, the kid's favorite candy, cookie and ice cream on a table along with a $20, $50 or a $100 bill and tell them they can only have one and make sure you record the event. If the kid takes the money or an eatable treat and then tries to negotiate another eatery treat from you then they are normal and have a healthy relationship with technology and understands what money is. If the kid takes the device without hesitation and leaves room and doesn't try to negotiate one of those sugary treats or money from you; shame on you—sure hope you've installed Kahoot on that device so they can take advantage of some digital learning until the battery dies. Not only are you contributing to your kid's electronic dependency; they have no concept of what money is and how delicious Oreo's and some homemade vanilla Blue Bell ice cream is; and the development of those critical social and interpersonal skills needed long after we're gone are not being mastered stunting their personal growth. Don't feel bad parents you're not alone. The next time you go out in public to a restaurant, retail store, the airport or grocery store (that's making groceries if you're from New Orleans); if you do what normal people have done for centuries and that is to simply be alert and look up; you will then notice 3 things: very few people are looking up, kids are staring into a device and not talking to anyone developing those social skills and the adults are sitting next to each other constantly on their device too, not even speaking. I can't tell you how many times I've been in the checkout line in the Grocery and Retail store and the person behind me is constantly on her phone with earbuds on, the kid in the buggy staring at a device and there's an expensive Louis Vuitton or Michael Kors purse left unattended in the same buggy. So; after I check out, I gesture to the cashier with the universal keep quiet sign and I wait for the employee to finish bagging her groceries, then I slowly pull her cart all the way out of line and just wait to see if she notices if her child, purse and cart is gone. It must have been a monumental conversation because almost 2 minutes elapsed before the cashier who was visibly annoyed and couldn't keep quiet a second longer said, "ma'am your child and purse is gone!" Can you imagine how the conversation with the authorities and husband would be if I actually would've abducted the kid, taken that expensive purse and drove away quietly in her $100k Tesla?…still smh…sbt We are all connected universally and digitally. The spirit of unity has been Corrupted. Why hasn't there been a cure for hatred or bigotry since there's a pill for everything else? Maybe it's hidden in those original Scribes or the Bible. Despite our perceived differences; whether spiritual or non-spiritual, it's time to Unite or call the Exopolitics expert Dr. Michael Salla to help us open up the imagination of future generations again to simply shoot for the stars. Grown up time out—with all of the advances in new technology, radar, satellites, Google Earth, the cloud, Mae East and Mae West, people are more tribal than ever. Will the gender war or technology be the tipping point? What say U….. Now put that digital device in selfie mode and hit the record button and repeat after me: "I pledge allegiance to my phone and the Internet States of America. And to the new Republic, in these lands; no more liberties, just cool emoji lies, iPhones, androids and free cells for all." Now make this book go viral ppl and tag me.

Glossary

Addiction – The compulsive psychological need for something, creating an abnormal dependency. hence, one of the many descriptions of behavioral addictions is the following: an irresistible urge, impulse or drive to repeatedly engage in an action and an inability to reduce or cease this behavioral

Amazon – is an American electronic commerce and cloud computing company based in Seattle, Washington— founded by Jeff Bezos on July 5, 1994; is the world's largest online retailer and a prominent cloud services provider. Amazon Web Services (AWS) is a comprehensive, evolving cloud computing platform;

Bout that Street Life Wannabees – Internet Thug who lives in the suburbs with a 2+ car garage with the Cool Mom with Tattoos who Twerks;

B4 – before; Beaucoup – Many or a lot;

Booty – valuable stolen goods, especially those seized in war

Bully – a person who uses strength or power to harm or intimidate those who are weaker to force him or her to do what one wants;

Chatbot – a robot who can mimic online language patterns of a human to better understand the way humans converse n relate;

Clap Back – responding to a criticism viciously; an acute comeback intended to place someone in much-needed check;

Cyber | Digital Cocaine – getting high on highly addictive Video games, smartphones and social media; can be harmful to the brain;

Cyberbullying – bullying that happens on social networking sites, that can be either physical, verbal or relational;

Digital Booty – highly valuable personal identifiable information stolen especially by stealth consent coercion via mobile apps; websites, terms and conditions to grant permission and access to be looted

Digital Drugs – more accurately called binaural beats, are sounds that are thought to be capable of changing brain wave patterns and inducing an altered state of consciousness similar to that effected by taking drugs or achieving a deep state of meditation;

Digital Pimps – Companies who own and profit from your (PII) and you don't;

Dope – not an intellectual giant; an addictive narcotic;

Drip Drip – Cool, hip, that's wassup and poppin;

Earbud Warriors – walks around detached from the real-world void of conversation; usually closed minded; constantly feels lonely, unenthused n shies away from making eye contact except when asking for $$$;

Ephemeral – lasting a very short time; short-lived; transitory;

Facebook (FB) – free social networking website that allows registered users to create profiles, upload photos, videos, send messages n keep in touch with friends, family n colleagues;

Fiend – person who is intensely interested in or fond of something;

Follower – person who follows another in regard to his or her ideas #bigdummy; a stalker or creeper in the real world

Fomo – Fear of Missing Out;

Go Ham – act crazy or ignorant without reason with little to no regard for anyone or anything regardless of time and place;

Google – a search engine that can be employed to find a variety of information such as websites, pictures, also uses a computer program called a 'web crawler' that looks at the billions of websites available on the

World Wide Web and examines their content to find 'keywords'. Then it indexes these to make the websites easier for the search engine to find;

Have to – hv2; Hella – A large amount of or a Number of;

Headshot – instant damage or a kill to a Fortnite video Skin

Hmu – hit me up or contact me

IAD – Internet Addiction Disorder;

IBO – Information Brain Overload

Instagram (IG) – online photo-sharing application, social network platform that allows users to edit and upload photos and short videos through a mobile app;

Internet Gangsta – One who uses the Internet as a front for acting like a tough person, gang member, never been in a real fight, usually because they are hoping to gain the respect that they lack in their real life;

Internet Troll – a person who starts quarrels or upsets people on the Internet to distract and sow discord by posting inflammatory and digressive, extraneous, or off-topic messages;

Internet Thug – person that avoids human confrontation and acts all hardcore on the internet but if you saw them in person you would laugh aka a Studio Gangsta in the music business;

Internet – Founded in 1969; a massive network of networks, a networking infrastructure that connects millions of computers together globally, forming a network in which any computer can communicate with any other computer connected to the Internet;

Internet Association – is a United States industry trade group based in Washington, D.C., which represents internet companies. It was founded in 2012 by several companies, including Google, Amazon, eBay, and Facebook, and is headed by the president and CEO Michael Beckerman. Their mission is to foster innovation, promote economic growth, and empower people through the free and open internet;

ISP – Internet Service Provider;

i-Dosing – finding an online dealer who can hook you up with "digital drugs" that get you high through donning headphones n listening to "music" – largely a droning noise – which the sites peddling the sounds promise will get you high;

Kahoot – is a game-based learning platform, used as educational technology in schools and other educational institutions

Kewl – Cool

Lame – out of touch with modern fads or trends; unsophisticated

Mobile Drug Addict – A low battery can cause right-out chaos and anxiety; lacks self-control; Spending time away from the phone makes you feel anxious or panicky, looking a bit like withdrawal and getting a text or instant message lift their moods; (loss of control) despite serious negative consequences to the person's physical, mental, social and/or financial well-being resulting in impulse control disorder cited in the ICD-11;

Mushed – The act of placing one's hand on another person's face and pushing the person,

Nomofobia – the anxiety or irrational fear of being without your mobile phone or being unable to use your phone for some reason, such as the absence of a signal or running out of minutes or battery power;

n = and; **n2** = into; **2u** = to you; **OG** – Original Gangster;

Pii – personal identifiable information;

Pull up – arriving solo or with your gang to fight or shoot down an enemy

Ratchet – someone that is mean, nasty, trifling and classless in a social setting; Sbt – sad but true

Selfie – photos that you take of yourself, usually with a mobile phone and often published using Social Media;

Shade – publicly criticize or express contempt for someone;

Smoke – new modern-day slang for there's an issue or beef; of the weed vernacular

Smoked – slang for dude got murdered, popped, capped, busted, copped, shot

Snapchat – mobile app that allows U2 to send/receive ephemeral photos n videos;

Social Media (SM) – websites and applications that enable users to create and share content or to participate in social networking;

Stacks = $1,000;

Social Media Zombie – not dead, but not quite alive either with their eyes glazed; who can't go to any game, restaurant or bar without telling their social networks where they are;

Stow Bought Sense – Every time you learn something you pay 4 it;

Swag – Stuff we all got, Cool n Confident;

Twitter – an online news and social networking site where people communicate in short messages called tweets, microblogging and to discover interesting people and companies online, opting to follow their tweets;

Twitter Thug – person that avoids human confrontation wholeheartedly and acts all hardcore on the internet— likes to clap back via a digital device

U = you, **Ur** = your; **U2** = you too;

Wiggity Whack – Something so horribly ridiculous;

World Wide Web – Since 1998 an information system on the Internet that allows documents to be connected to other documents by hypertext links, enabling the user to search for information by moving from one document to another;

Zombie – the living dead that is capable of movement but not of rational thought;

Slang Terms from 1960 – 2002: Ask Siri, Alexa, Google or people with Real Gray Hair who listened to rock n roll, funk and real hip hop music

To all of the Future Generations out there in middle and high school; it's perfectly okay if you don't know what you wanna do later on in life career wise. Your career will ultimately find you no matter what direction your protectors and mentors try to set your life's GPS too; because the life challenges that your parents and grandparents faced long ago will one day be staring at you in real-time, so enjoy every moment of your age right now. To all future and current women: life is a game of Chess; so don't ever voluntarily give up your Queen card because the Queen is the most powerful piece on the chess board and always wills the power behind the throne. To all future and current men in the #me2 era; don't ever apologize for being who you are and walk and talk with confidence. Now everyone young and old—get off of social media and the internet and go live your real life and enjoy your Treasures!

Do U hv Real World Courage like this 6yr old?

www.ingramcontent.com/pod-product-compliance
Lightning Source LLC
Chambersburg PA
CBHW041132200526
45172CB00018B/101